AF712084

Cornell University Library
Ithaca, New York

BOUGHT WITH THE INCOME OF THE
SAGE ENDOWMENT FUND
THE GIFT OF
HENRY W. SAGE
1891

Cornell University Library
arV18626

Hindu achievements in exact science;

3 1924 031 220 126

HINDU ACHIEVEMENTS IN EXACT SCIENCE

HINDU ACHIEVEMENTS IN EXACT SCIENCE

A Study in the History of Scientific Development

BY

BENOY KUMAR SARKAR

Professor, National Council of Education, Bengal
Author of "The Positive Background of Hindu Sociology," "Chinese Religion through Hindu Eyes," "The Science of History and the Hope of Mankind," etc.

LONGMANS, GREEN AND CO.
FOURTH AVENUE & 30TH ST., NEW YORK
39 PATERNOSTER ROW, LONDON
BOMBAY, CALCUTTA, AND MADRAS
1918

Copyright, 1918
BY
LONGMANS, GREEN AND CO.

PRESS OF
BRAUNWORTH & CO.
BOOK MANUFACTURERS
BROOKLYN, N. Y.

PREFACE

THE main object of this little book is to furnish some of the chronological links and logical affinities between the scientific investigations of the Hindus and those of the Greeks, Chinese, and Saracens. Details relating to the migration of discoveries have been generally avoided, as they require a treatment more technical than the present scope and space admit. Nor have all the achievements of the Hindus in any branch of science been treated in an exhaustive manner.

It has been sought to present a comprehensive, though very brief account of the entire scientific work of ancient and mediæval India in the perspective of developments in other lands. From the standpoint of modern science a great part of all that is described here is too elementary to have more than an anthropological interest. If, however, the facts of Hindu and Chinese science were made available in more extensive volumes than has yet been done, the students of comparative culture-history would find that the tendencies of the Oriental mind have not been essentially distinct from those of the Occidental.

The works that have been frequently consulted are given in the Bibliography. Special mention must be made of the writings of Professors Nalinbehari Mitra and Brajendranath Seal. For some of the Sanskrit literature on Hindu science I have made use of my "Positive Background of Hindu Sociology."

I am indebted to the authorities of Harvard University for the privilege of using its library.

BENOY KUMAR SARKAR

Cambridge, Mass.
April 15 1917

CONTENTS

BIBLIOGRAPHY

N.B.—The figures in the text refer to the numbers on this list.

1. Aristotle—"On the parts and progressive motion of animals; on invisible lines; true arithmetic of infinites, etc." (translated by Taylor), London, 1810.
2. Bryant—"History of Astronomy," London, 1907.
3. Burgess—"Surya-Siddhanta" (Hindu astronomy), American Oriental Society, New Haven, 1860.
4. Cajori—"History of Mathematics," New York, 1909.
5. Chatterji—"The Economic Botany of India," District Council of National Education, Malda, Bengal, India, 1910.
6. Clements—"Introduction to the Study of Indian Music," London, 1913.
7. Colebrooke—"Algebra with Arithmetic and Mensuration from the Sanskrit of Brahmagupta and Bhaskara," London, 1817.
8. Colebrooke—"Essays," Vol. 2, London, 1873.
9. Cressy—"Discoveries and Inventions of the Twentieth Century," London, 1914.
10. Deval—"The Hindu Musical Scale and the Twenty-two Shrutees," Aryabhusan Press, Poona, India, 1910.
11. Dutt—"The Materia Medica of the Hindus," Calcutta, 1900.

12. Elson—"The Book of Musical Knowledge," Boston, 1915.

13. Encyclopædia Britannica, eleventh edition (article on Mineralogy).

14. Fox-Strangways—"The Music of Hindostan," Oxford, 1914.

15. Freind—"History of Physick from the Time of Galen to the Beginning of the Sixteenth Century," London, 1727.

16. Garrison—"History of Medicine," Philadelphia, 1914.

17. Geddes and Thomson—"Sex," London, 1914.

18. Gibbon—"History of the Saracen Empire" (reprint of the fiftieth, fifty-first, and fifty-second chapters of the monumental History), London, 1870.

19. Gondal—"History of Aryan Medical Science," London, 1896.

20. Greene—"Landmarks of Botanical History," Washington, D. C., 1909.

21. Heath—"Aristarchus of Samos, the Ancient Copernicus," Oxford, 1913.

22. Heath—"Diophantus of Alexandria," Cambridge, 1910.

23. Hoernle—"Medicine of Ancient India," Oxford, 1907.

24. Kaye—"Indian Mathematics," Calcutta, 1915. (This book must not be read without Mitra's criticism on it in "Hindu Mathematics.")

25. Leighton—"Embryology," London, 1914.

26. Lewes—"Aristotle: A Chapter in the History of Science," London, 1864.

27. Libby—"History of Science," Boston, 1917.

28. Mehta—"Ayurvedic System of Medicine," Bombay, 1913.

29. Meryon—"History of Medicine," London, 1861.

30. Merz—"History of European Thought in the Nineteenth Century," Edinburgh, 1896.

31. Meyer—"History of Chemistry" (translated by McGowan from German), London, 1891.

32. Mikami—"Development of Mathematics in China and Japan," Leipzig, 1913.

33. Mitra—"Hindu Mathematics," The Modern Review Office, Calcutta, 1916.

34. Mookerji—"History of Indian Shipping," London, 1912.

35. Muir—"Heroes of Science: Chemists," London, 1883.

36. Muir—"The Story of Alchemy," London, 1902.

37. Müller—"India: What Can it Teach Us?" London, 1883.

38. "The Nineteenth Century: A Record of Progress," New York, 1901.

39. Oppert—"Fire-arms and Weapons of the Hindus," Madras, 1880.

40. Pettigrew—"Superstitions Connected with the History and Practice of Medicine and Surgery," London, 1844.

41. Playfair—"Progress of Mathematical and Physical Science," Edinburgh (undated).

42. Pliny—"Natural History" (translated by Bostock and Riley), London, 1855–57.

43. Powell—"History of Natural Philosophy," London, 1837.

44. Ray—"History of Hindu Chemistry," Calcutta, 1902, 1909.

45. Ray-Lankester—Article on Zoology in the Encyclopædia Britannica.

46. Reid—"Laws of Heredity," London, 1910.

47. Rosen—"The Algebra of Mohammed ben Musa," London, 1831.

48. Routledge—"Discoveries and Inventions of the Nineteenth Century," London, 1899.

49. Roy—"The Astronomy and Astronomers of the Hindus" (in Bengali), Calcutta, 1903.

50. Roy—"Ratna Pariksa" (Examination of Gems, in Bengali), Calcutta, 1904.

51. Royle—"Antiquity of Hindu Medicine," London, 1837.

52. Royle—"The Arts and Manufactures of India" (Lectures on the Results of the Great Exhibition of 1851), London, 1852.

53. Sachau—"Alberuni's India," London, 1910.

54. Sachs—"History of Botany (1530–1860)," Oxford, 1890.

55. Sarkar—"The Positive Background of Hindu Sociology," Vol. I., Panini Office, Allahabad, India, 1914.

56. Sastri—"The Educative Influence of Sanskrit Literature," Benares, India, 1916.

57. Schoff—"The Periplus of the Erythræan Sea" (Travel and Trade in the Indian Ocean by an Egyptian Greek Merchant of the First Century A.D.), London, 1912.

58. Scott—"History of the Moorish Empire in Europe," Philadelphia, 1904.

59. Seal—"The Positive Sciences of the Ancient Hindus," London, 1916.

60. Sinha—"The Vaishesika Sutras of Kanada" (Hindu Democritus), Panini Office, Allahabad, 1911.

61. Smith and Karpinski—"Hindu-Arabic Numerals," Boston, 1911.

62. Smith and Mikami—"Japanese Mathematics," Chicago, 1914.

63. Tagore—"Hindu Music from Various Authors," Calcutta, 1875.

64. Takakusu—"Itsing: Records of the Buddhist Religion (671–95)," Oxford, 1896.

65. Thibaut—"On the Sulva-sutras" (Journal of the Asiatic Society of Bengal, Vol. XLIV, Reprint, Calcutta, 1875).

66. Thompson, D'A. W.—"On Aristotle as Biologist."

67. Thomson, J. A.—"Heredity," London, 1912.

68. Thomson, T.—"History of Chemistry," London, 1831.

69. Thomson, T.—"Progress of Physical Science," New York, 1843.

70. Thorpe—"History of Chemistry," New York, 1910.

71. Tilden—"Progress of Scientific Chemistry," London, 1899.

72. Venable—"Short History of Chemistry," Boston, 1894.

73. Wallace—"The Wonderful Century," New York, 1898.

74. Wallace and Others—"The Progress of the Century," New York, 1901.

75. Werner—"Chinese Sociology," London, 1910.

76. Whewell—"The History of the Inductive Sciences," New York, 1858.

77. Williams—"The Middle Kingdom" (China), New York, 1914.

78. Wilson—"Essays on Sanskrit Literature," Vol. I, London, 1864.

79. Wise—"Commentary on the Hindu System of Medicine," Calcutta, 1845.

HINDU ACHIEVEMENTS

HISTORICAL PERSPECTIVE

INVESTIGATIONS in radioactivity since 1896 have effected a marvellous revolution in our knowledge of Energy. The ultimate atoms of matter are now believed to possess "sufficient potential energy to supply the uttermost ambitions of the race for cosmical epochs of time."

Speaking of these new discoveries in connection with radioactivity, Professor Soddy remarks in his "Matter and Energy":

"It is possible to look forward to a time, which may await the world, when this grimy age of fuel will seem as truly a beginning of the mastery of energy as the rude stone age of palaeolithic man now appears as the beginning of the mastery of matter."

This optimism seems almost to out-Bacon Bacon's prophecy in the "Novum Organum" (1621) relating to the wonderful achievements he expected from a "new birth of science." The "new birth" was inevitable, he declared, "if any one of ripe age, unimpaired senses, and well-purged mind, apply himself anew to experience and particulars."

Becquerel's discovery of radioactive substances is thus a quarter less than three hundred years from Bacon's first advocacy of experimental and inductive methods. The "long and barren period" between the scientific activity of ancient Greece and that of modern Europe, described by Whewell as the "stationary period of science," was drawing to a close in Bacon's time. The age was, however, yet "dark" enough to be condemned by him in the following words:

"The lectures and exercises there (at the universities) are so ordered that to think or speculate on anything out of the common way can hardly occur to any man. Thus it happens that human knowledge, as we have it, is a mere medley and ill-digested mass, made up of much credulity and much accident, and also of childish notions which we at first imbibed."

Positive Science is but three hundred years old. It is necessary to remember this picture of the intellectual condition of Europe at the beginning of the seventeenth century in every historical survey of the "exact" sciences (whether deductive-mathematical or inductive-physical), as well as in every comparative estimate of the credit for their growth and development due to the different nations of the world.

Hindu investigations in exact science, as briefly

summarized here, come down to about 1200 A.D. Strictly speaking, they cover the period from the "Atharva Veda" (*c* 800 B.C.), one of the Hindu Scriptures, to Bhaskaracharya (*c* 1150), the mathematician; or rather to the middle of the fourteenth century, represented by Madhavacharya, the compiler of "The Sixteen Systems of Philosophy" (1331), Gunaratna (1350), the logician, "Rasa-ratna-samuchchaya," the work on chemistry, and Madanapala, author of the "Materia Medica" (1374) named after himself.

We are living to-day in the midst of the discoveries and inventions of the last few years of the twentieth century, e.g., those described in Cressy's volume. To the moderns, therefore, the whole science of the Hindus exhibited here belongs to what may be truly called the pre-scientific epoch of the history of science. Its worth should, however, be estimated in the light of the parallel developments among their contemporaries, the Greeks, the Chinese, the Graeco-Romans, the Saracens, and the mediæval Europeans.

Whewell, according to whom the scientific inquiries of the ancients and mediævals "led to no truths of real or permanent value," passes the following summary and sweeping judgment on all these nations:

"Almost the whole career of the Greek schools of philosophy, of the schoolmen of Europe in the Middle Ages, of the Arabian and Indian philosophers, shows that we may have extreme ingenuity and subtlety, invention and connection, demonstration and method; and yet out of these no physical science may be developed. We may obtain by such means logic and metaphysics, even geometry and algebra; but out of such materials we shall never form optics and mechanics, chemistry, and physiology."

Further, "the whole mass of Greek philosophy shrinks into an almost imperceptible compass, when viewed with reference to the progress of physical knowledge." "The sequel of the ambitious hopes, the vast schemes, the confident undertakings of the philosophers of ancient Greece was an entire failure in the physical knowledge."[76]

While accepting for general guidance the above estimate of Whewell regarding the ancients and mediævals, the student of Culture-history would find the following noteworthy points in a survey of the world's positive sciences from the Hindu angle

1. The "pure" mathematics of the Hindus was, on the whole, not only in advance of that of the Greeks, but anticipated in some remarkable instances the European discoveries of the sixteenth, seventeenth, and eighteenth centuries. That mathematics is the basis of the mathematical science known to modern mankind.

2. Like the other races, the Hindus also may be taken to have failed to make any epoch-making discoveries of fundamental "laws," planetary, inorganic, or organic, if judged by the generalizations of to-day. But some of their investigations were solid achieve-

ments in positive knowledge, viz., in materia medica, therapeutics, anatomy, embryology, metallurgy, chemistry, physics, and descriptive zoology. And in these also, generally speaking, Hindu inquiries were not less, if not more definite, exact, and fruitful than the Greek and mediæval-European.

3. Hindu investigations helped forward the scientific developments of mankind through China (and Japan) on the east and the Saracens on the west of India, and this both in theoretical inquiries and industrial arts.

4. Since the publication of Gibbon's monumental history, the historians of the sciences have given credit to the Saracens for their services in the development of European thought. Much of this credit, however, is really due to the Hindus. Saracen mathematics, chemistry, and medicine were mostly direct borrowings from Hindu masters. The Greek factor in Saracen culture is known to every modern scholar; the Hindu factor remains yet to be generally recognized. That recognition would at once establish India's contributions to Europe.

5. Every attempt on the part of modern scholars to trace the Hellenic or Hellenistic sources of Hindu learning has been practically a failure.

6. But, like every other race, the Hindus also got their art of writing from the Phœnicians. Besides, the Hindus may have derived some inspiration from Greece in astronomy as admitted by their own scientists, e.g., by Varahamihira (505–587 A.D.).

India's indebtedness to foreign peoples for the main body of her culture is virtually nil.

7. Hindu intellect has thus independently appreciated the dignity of objective facts, devised the methods of observation and experiment, elaborated the machinery of logical analysis and truth investigation, attacked the external universe as a system of secrets to be unravelled, and wrung out of Nature the knowledge which constitutes the foundations of science.

8. The claims of the Hindus to be regarded as pioneers of science and contributors to exact, positive, and material culture rest, therefore, in all respects, on the same footing as those of the Greeks, in quality, quantity, and variety. An absolute superiority cannot be claimed for either, nor can any fundamental difference in "weltanschauung," mental outlook, or angle of vision be demonstrated between these two races.

It has been remarked above that the age of experimental and inductive science is about three hundred years. It is this period that has established the cultural superiority of the Occident over the Orient. But this epoch of "superiority" need be analyzed a little more closely.[55]

Neither the laws of motion and gravitation (of the latter half of the seventeenth century), nor the birth of the sciences of modern chemistry and electricity during the latter half of the eighteenth could or did produce the superiority in any significant sense. There was hardly any difference [48] between Europe and Asia at the time of the French Revolution (1789). The real and only cause of the parting of ways between the East and the West, nay, between the mediæval and the modern, was the discovery of steam, or rather its application to production and transportation. The steam engine effected an industrial revolution during the first three decades of the nineteenth century. It is this revolution which has ushered in the "modernism" of the modern world in social institutions, science, and philosophy,[30] as well as brought about the supremacy of Eur-America over Asia.

The year 1815 may be conveniently taken to be the year I of this modernism, as with the fall of Napoleon it marks also the beginning of a new era in world-politics, practically the era in which we still live. The difference between the Hindu and the Eur-American, or between the East and the West, is a real difference to-day. But it is not a difference in mentality or "ideals" or so-called race-genius. It is a difference of one century, the "wonderful century," in a more comprehensive sense than Wallace gives to it.

I

ARITHMETIC

A GENERAL idea of the achievements of the Hindu brain may be had from the following remarks of Cajori:

"It is remarkable to what extent Indian mathematics enter into the science of our time. Both the form and the spirit of the arithmetic and algebra of modern times are essentially Indian and not Grecian. Think of that most perfect of mathematical symbolisms, the Hindu notation, think of the Indian arithmetical operations nearly as perfect as our own, think of their algebraic methods, and then judge whether the Brahmins on the banks of the Ganges are not entitled to some credit. Unfortunately some of the most brilliant of the Hindu discoveries in indeterminate analysis reached Europe too late to exert influence they would have exerted, had they come two or three centuries earlier."

As De Morgan admits, "Hindu arithmetic is greatly superior to any which the Greeks had. Indian arithmetic is that which we now use." [33]

The Hindus were the greatest calculators of antiquity. They could raise the numbers to various powers. The extraction of square or cube root was a child's play to them.

The two foundations of arithmetic were discovered by the Hindus:

(1) The symbols of numbers, or numerals as they are called, and

(2) The decimal system of notation.

Numerals have been in use in India since at least the third century B.C. They were employed in the Minor Rock Edicts of Asoka the Great (256 B.C.). In modern times the numerals are wrongly known as "Arabic" because the European nations got them from their Saracen (Arab) teachers.[61]

The decimal system was known to Aryabhata (A.D. 476) and Brahmagupta (A.D. 598–660), and fully described by Bhaskaracharya (1114). In Subandhu's "Vasavadatta," a Sanskrit prose romance (A.D. 550–606?) the stars are described as zero. In "Vyasa-bhasya," also, the system is referred to. The transformation of substance in chemical fusion through the "unequal distribution of forces" is there illustrated by a mathematical analogy: "Even as the same figure '1' stands for a hundred in the place of hundred, for ten in the place of ten, and for a unit in the place of a unit." The "Vyasa-bhasya" cannot have been composed later than the sixth century A.D. The decimal system was therefore known to the Hindus long before its appearance in the writings of the Arabs or Græco-Syrians.[59]

The Saracens learnt from the Hindus both the system of numeration and the method of computation. Even in the time of Caliph Walid (705–15) the Saracens had to depend on alphabetical symbols. They had no figures for numbers yet. A Hindu scientific mission reached Mansur's court from Sindh in 773. This introduced the Moslems to Hindu astronomical tables. The Saracen astronomical work thus compiled was abridged by Musa, the Librarian of Caliph Mamun (813–33). "And he studied and communicated to his countrymen the Indian compendious method of computation, i.e., their arithmetic, and their analytic calculus." [7]

This was the first introduction of the decimal system among the Saracens (830). They have ever since acknowledged their debt to the Hindus. Alberuni (1033) wrote: "The numeral signs which we use are derived from the finest forms of the Hindu signs."

It was probably in the twelfth century that the Europeans learnt this Hindu science from their Saracen masters. Leonardo of Pisa,[61] an Italian merchant, was educated in Barbary, and thus became acquainted with the so-called Arabic numerals and Musa's work on Algebra based on the Sanskrit. In 1202 was published his "Liber Abbaci." This was the beginning of modern arithmetic in Europe. The pioneering work may have been done by Gerbert, a Frenchman, who learnt the Hindu system from the Mohammedan teachers at Cordova in Spain (*c* 970–80).[69] Musa, the distinguished Moslem mathematician, was thus the connecting link between the algebra and arithmetic of the Hindus and the mediæval European mathematics.

At the commencement of the Christian era, the Chinese "adopted the decimal system of notation introduced by the Buddhists, and changed their ancient custom of writing figures from top to bottom for the Indian custom of from left to right." [75]

II

ALGEBRA

ALGEBRA is a Hindu science in spite of the Arabic name. Cajori suspects that Diophantus (A.D. 360), the first Greek algebraist, got the first glimpses of algebraic knowledge from India. According to Heath, the Europeans were anticipated by the Hindus in the symbolic form of algebra. According to De Morgan, the work of Diophantus is hardly algebraic in the sense in which that term can be applied to the science of India.[33] According to Hankel, the Hindus are the real inventors of algebra if we define algebra as the application of arithmetical operations to both rational and irrational numbers or magnitudes.[4]

The mathematician who systematized the earlier algebraic knowledge of the Hindus and thus became the founder of a new science is Aryabhata (born A.D. 476 at Pataliputra on the Ganges in Eastern India). He was thus over a century later than Diophantus; but Smith proves that neither in methods nor in achievements could the Greek be the inspirer of the Hindu.[33]

The points in which the Hindu algebra appears particularly distinguished from the Greek are thus enumerated by Colebrooke:

1. A better and more comprehensive algorithm.
2. The management of equations involving more than one unknown term. (This adds to the two classes noticed by the Saracens, viz., simple and compound.)
3. The resolution of equations of a higher order, in which if they achieved little, they had at least the merit of the attempt, and anticipated a modern discovery in the solution of biquadratics.
4. General methods for the solution of indeterminate problems of first and second degrees, in which they went far beyond Diophantus, and anticipated the discoveries of modern algebraists.
5. Application of algebra to astronomical investigation and geometrical demonstration, in which also they hit upon some methods which have been reinvented in later times.

It was thus not a "primitive" algebra that the Hindus developed. The achievements of Indian algebra from the fifth to twelfth century A.D. have in some cases anticipated the discoveries of the seventeenth and eighteenth centuries in Europe. Modern algebraists have thus only re-discovered the already known truths.

The Hindu algebra of this period was the principal feeder of Saracen algebra through Yakub and Musa, and indirectly influenced to a certain extent mediæval European mathematics. It may have fostered the development of mathematics in China also, and through that, of Japan. According to Williams, the Hindu processes in algebra were known to the mathematicians of the Chinese empire, "and are still studied" in the Middle Kingdom "though all intellectual intercourse between the two countries has long ceased."

The progress of Hindu algebra (mainly in Southern India) after Bhaskara (twelfth century) was, as Seal suggests, parallel to the development in China and Japan. But this is a subject that awaits further research.

The Hindu discoveries in algebra may be thus summarized from the recent investigations of Nalinbehari Mitra:

1. The idea of an absolutely negative quantity.
2. The first exposition of the complete solution of the quadratic equation (Brahmagupta A.D 598–660)
3. Rules for finding permutations and combinations Bhaskara, born 1114). These were unknown to the Greeks.
4. Indeterminate equations: "The glory of having invented general methods in this most subtle branch of mathematics belongs to the Indians." [4]
5. Indeterminate equations of the second degree.

In the light of Comparative Chronology, these discoveries are remarkable evidences of the fecundity of the Hindu brain in "exact" science. The three great anticipations of modern algebra are enumerated and appreciated by Colebrooke in the following terms:

1. The demonstration of the noted proposition of Pythagoras concerning the square of the base of a rectangular

triangle, equal to the squares of the two legs containing a right angle. The demonstration is given in two ways in Bhaskara's algebra (twelfth century). The first of them is the same which is delivered by Wallis (1616–1703) in his treatise on angular sections, and as far as appears, then given for the first time.

2. The general solution of indeterminate problems of the first degree. It was first given among moderns by Bachet de Meziriac in 1624.

3. Solution of indeterminate problems of the second degree . . . a discovery which among the moderns was reserved for Euler (1707–83). To him among the moderns we owe the remark which the Hindus had made more than a thousand years ago, that the problem was requisite to find all the possible solutions of equations of this sort.

Bhaskara invented the art of placing the numerator over the denominator in a fraction. He invented also $\sqrt{}$ (the radical sign). This was not known in Europe before Chuquet and Rudolff in the sixteenth century.

Bhaskara also proved the following:

$$x+0=x; \quad 0^2=0; \quad \sqrt{0}=0; \quad x\div 0=\infty$$

III

GEOMETRY

The earliest geometry of the Hindus is to be found in the "Sulva-sutras"[65] of Baudhayana and Apastamba. In these treatises, which form parts of the Vedic literature, we get the application of mathematical knowledge to the exigencies of religious life, sacrifices, rituals, construction of altars, etc.

At this stage Hindu geometry was quite independent of Greek influence. The following are some of the problems,[33] which were solved by the mathematicians of the Vedic cycle:

1. The so-called Pythagorean theorem: The square on the hypotenuse of a right-angled triangle is equal to the sum of the squares on the other two sides.
2. Construction of squares equal to the sum or difference of two squares;
3. Conversion of oblongs into squares, and *vice versa;*
4. Drawing of a perpendicular to a given straight line at a given point of it;
5. Construction of lengths equal to quadratic surds: The approximate value of $\sqrt{\ }$
6. Circling of squares;
7. Squaring of circles,—"that rock upon which so many

reputations have been destroyed," both in the East and West. The earliest Hindus got $\pi = 3.0044$;

8. Construction of successive larger squares from smaller ones by addition;
9. Determination of the area of a trapezium, of an isosceles trapezium, at any rate, when the lengths of its parallel sides and the distance between them are known.

The oldest geometrical efforts of the Hindus were not entirely empiric. They doubtless "reasoned out all or most of their discoveries." [4] These could not have been inspired by the Greeks.[33]

We find Aryabhata (A.D. 476) solving the following among other problems, viz., the determination of:

1. The area of a triangle;
2. The area of a circle;
3. The area of a trapezium;
4. The distance of the point of intersection of the diagonals of a trapezium from either of the parallel sides;
5. The length of the radius of a circle.

Aryabhata gave also the accurate value of $\pi (= \frac{62832}{20000})$ and the area of the circle as πr^2. The Saracens learnt this from the Hindus. Probably Yakub (eighth century) was the first to get it when the astronomical tables were imported to Bagdad from India. The correct value of π was not known in Europe before Purbach (1423–61).

At this stage also Hindu geometricians were not indebted to the Greeks. Their independence is thus argued by Mitra:

"Euclid and his school never meddled with *logistics* which was practically abandoned as hopeless after the time of Apollonius, while the Indian mathematician's turn of mind was nothing if it was not directed to practical computations. The fact that the Indians took the chord of a smaller circular arc as equivalent in length to the arc—a step which no sane Greek mathematician with a free conscience would have even dreamt of taking—ought to settle once for all the question of the dependence of Indian geometry on Greek geometry."

Fresh contributions [33] to geometry were made by Brahmagupta (598–660); viz., those relating to

1. The construction of right-angled triangles with rational sides;
2. The various properties of right-angled triangles;
3. The area of a cyclic quadrilateral;
4. The properties of isosceles trapezium;
5. The properties of cyclic quadrilateral;
6. The properties of circles; Brahmagupta gave the rules (1) for finding the diameter of a circle when the height and chord of a segment of it are given, and (2) for finding the area of a segment of a circle. The first rule in the form given by the Hindu was not known in Greece. Musa (830) learnt both these rules from Brahmagupta's works.
7. The volume of a cone as one-third the volume of the cylinder;
8. The volume of a pyramid as one-third the volume of the prism;
9. The volume of a cavity of uniform bore (prismatic or cylindrical).

Bhaskara (1114) summarized and methodized the results of all previous investigators, e.g., Lata, Aryabhata, Lalla (499), Varahamihira (505), Brahmagupta, Shreedhara (853), Mahavira (850), Aryabhata the Younger (970), and Utpala (970).

Bhaskara took care to explain that though Aryabhata and others knew the exact value of π, yet some later mathematicians took approximate values only for convenience of calculation. "It is not that they did not know." Thus Brahmagupta took $\pi=3$ roughly (or $\sqrt{10}$ closely) "for lessening the labor of calculation."

Among Bhaskara's original contributions may be mentioned the fact that he gave two proofs of the so-called Pythagorean theorem. One of them was "unknown in Europe till Wallis (1616–1793) rediscovered it."[4]

It must be admitted that though Hindu geometricians achieved much the same results as the Greek, they did not attain the excellence of Euclid (*c* 306–293 B.C.) in method and system.

IV

TRIGONOMETRY

HINDU trigonometry was in advance of the Greek in certain particulars. The Hindus anticipated modern trigonometry also in a few points.

The mathematicians of India devised (1) the table of sines, and (2) the table of versed sines. The term "sine" is an Arabic corruption from Sanskrit "shinjini."

The use of sines was unknown to the Greeks. They calculated by the help of the chords.

The Hindu table of sines exhibits them to every twenty-fourth part of the quadrant, the table of versed sines does the same. In each, the sine or versed sine is expressed in minutes of the circumference, neglecting fractions.

The rule for the computation of the sines indicates the method of computing a table by means of their second differences. This is a considerable refinement in calculation, and was first practised in modern times by the English mathematician Briggs (1556–1631).

The astronomical tables of the Hindus prove that they were acquainted with the principal theorems of spherical trigonometry.

V

CO-ORDINATE GEOMETRY

VACHASPATI (A.D. 850), the Doctor of Nyaya (logic), anticipated in a rudimentary way the principle of co-ordinate (solid) geometry eight centuries before Descartes (1596–1650).

Vachaspati's claims are thus presented by Seal:

"To conceive position in space, Vachaspati takes three axes, one proceeding from the point of sunrise in the horizon to that of sunset, on any particular day (roughly speaking, from the east to the west); a second bisecting this line at right angles on the horizontal plane (roughly speaking, from the north to the south); and the third proceeding from the point of their section up to the meridian section of the sun on that day (roughly speaking, up and down). The position of any point in space, relatively to another point, may now be given by measuring distances along these three directions, i.e., by arranging in a numerical series the intervening points of contact, the lesser distance being that which comes earlier in this series, and the greater which comes later. The position of any single atom in space with reference to another may be indicated in this way with reference to the three axes.

"But this gives only a geometrical analysis of the conception of three-dimensioned space, though it must be admitted in all fairness that by dint of clear thinking it anticipates in a rudimentary manner the foundations of solid (co-ordinate) geometry."

VI

DIFFERENTIAL CALCULUS

BHASKARACHARYA anticipated Newton (1642–1727) by over five hundred years (1) in the discovery of the principles of differential calculus and (2) in its application to astronomical problems and computations.

According to Spottiswoode, the formula established by Bhaskara and "the method of establishing it bear a strong analogy to the corresponding process in modern mathematical astronomy,"[59] viz., the determination of the differential of the planet's magnitude.

According to Bapudeva Shastri,[59] Bhaskara's conception of instantaneous motion and the method of determining it indicate that he was acquainted with the principle of differential calculus.

According to Seal, Bhaskara's claim is indeed far stronger than Archimedes' to the conception of a rudimentary process of integration.

"Bhaskara, in computing the instantaneous motion of a planet compares its successive positions, and regards its motion as constant during the interval (which of course cannot be greater than a *truti* of time, i.e., $\frac{1}{33750}$th part of a second, though it may be infinitely less)."

This process is not only "analogous to, but virtually identical with, that of the differential calculus." As Spottiswoode remarks, mathematicians in Europe will be surprised to hear of the existence of such a process in the age of Bhaskara (twelfth century).

The claim for Bhaskara is, however, limited to the historically imperfect form of the calculus. Bhaskara does not specifically state that the method of the calculus is only approximative. But it must be remembered that the conception of limit and the computation of errors came late in the history of the calculi of fluxions and infinitesimals. For the rest, Bhaskara introduces his computation expressly as a "correction" of Brahmagupta's rough simplification.[59]

Further, Bhaskara's formula for the computation of a table of sines also implies his use of the principle of differential calculus.

VII

KINETICS

THE Hindus analyzed the concept of motion from terrestrial and planetary observations. To a certain extent they approached, though, strictly speaking, they did not anticipate, modern mechanics.

(1) Gravity: In astronomical works, e.g., of Aryabhata, Brahmagupta, and Bhaskara, the movement of a falling body is known to be caused by gravity. They ascribed gravity to the attraction exercised by the earth on a material body. But Newton's "law" of gravitation was not anticipated,

(2) Acceleration: Motion was conceived as a change of place in a particle and incapable of producing another motion; but "the pressure, impact, or other force which produces the first motion produces through that motion a *samskara* or persistent tendency to motion (*vega*), which is the cause of continued motion in a straight line, i.e., in the direction of the first motion." [59] A series of *samskaras* each generating the one that succeeded it was also conceived. Acceleration is thus logically implied in the writings of Udyotakara, the Doctor of Nyaya (logic).

(3) Law of Motion: The force of *samskara* (or persistent tendency to motion, i.e., *vega*) was known to diminish by doing work against a counteracting force; and when the *samskara* is in this way entirely destroyed, the moving body was known to come to a rest. Thus "*vega* corresponds to inertia in some respects, and to momentum (impressed motion) in others. This is the nearest approach to Newton's First Law of Motion,"[59] in the writings of Shamkara Mishra, the Doctor of Vaishesika (atomistic, Democritean) philosophy.

(4) Accelerated motion of falling bodies: Prashastapada (fourth century, A.D.), the Doctor of Vaishesika philosophy, believed that in the case of a falling body there is the composition of gravity with *vega* (momentum) acting in the same direction from the second instant onwards. It is as if the two motions coalesced and resulted in one. "Here is a good foundation laid for the explanation of the accelerated motion of falling bodies; but Galileo's discovery was not anticipated as Galileo's observations and measurements of motion are wanting." [59]

Scientifically considered, Hindu ideas on statics do not seem to have made much progress. It is interesting to observe that among the Greeks statics was more developed than dynamics. This is the exact opposite of the state of investigation in India, where motion was probably understood better than rest.

Thus the Hindus do not appear to have discovered the two celebrated principles of Archimedes (287–212 B.C.), viz.,

(1) That relating to the equilibrium of bodies and centre of gravity as determined by the balance—the first principle of Statics: Those bodies are of equal weight which balance each other at equal arms of a straight lever.

(2) That relating to the floating of bodies on liquids and the determination of specific gravity,—the first principle of Hydrostatics: A solid body, when immersed in a liquid, loses a portion of its weight equal to the weight of the liquid it displaces.

VIII

ASTRONOMY

Astronomical lore is probably as old as mankind. Elementary knowledge about the celestial bodies and meteorological phenomena is common to the races of antiquity, e.g., Chaldæans, Egyptians, Chinese, Hindus, and Greeks, as well as to all primitive races of men. That, however, is not to be regarded as forming the science of astronomy, unless the epoch of mere observation be lifted up to the level of an epoch of science.

The cultivation of astronomy, *as science*, after it began as such, did not make less progress among the Hindus than among the Greeks under Hipparchus (*c* 150 B.C.) and Ptolemy (A.D. 139).

1. Lunar zodiac: The earliest astronomy of the Hindus is believed to have been borrowed from the Babylonians. This consisted in the conception of the lunar zodiac with twenty-seven *naksatras* (constellations). But this elementary division of the sky, suggested by the passage of the moon from any point back to the same point, may have been original to the Hindu priests, as Colebrooke and Max Müller believe. The Saracens, however, learned

their *manzil* (twenty-eight constellations) from the Hindus in the eighth century.

2. Dodecameries: Aryabhata (A.D. 476) knew of the division of the heavens into twelve equal portions or "dodecameries." This zodiacal division came down from the Babylonians to the Greeks about 700 B.C. (?). But it was only by the first century B.C. that the Greeks had twelve separate signs for the twelve divisions. Aryabhata named the twelve divisions by words of the same import, and represented them by the figures of the same animals, as the Greeks. The Hindu zodiac, if it is foreign at all, seems thus to have been derived from the Greek, rather than from the Babylonian.

3. Rotation, 4. Eclipses: Aryabhata knew the truth that the earth revolves on its axis. The true cause of solar and lunar eclipses also was explained by him.

5. Epicycles: The hypothesis of the epicycles in accounting for the motions of the planets and in calculating their true places was the greatest generalization of Hipparchus. This was discovered by the Hindus also. But according to Burgess, "the difference in the develpoment of this theory in the Greek and the Hindu systems of astronomy precludes the idea that one of these peoples derived more than a hint respecting it from the other."

6. Annual precession of the equinoxes, 7. Relative size of the sun and the moon as compared with the earth, 8. The greatest equation of the centre for the

sun: With regard to these calculations, the Hindus "are more nearly correct than the Greeks."[3]

9. Times of the revolutions of the planets: With regard to these, the Hindus are "very nearly as correct" as the Greeks, "it appearing from a comparative view of the sidereal revolutions of the planets that the Hindus are most nearly correct in four items, Ptolemy in six."[3]

10. The determination of the lunar constants entering into the calculation of lunar periods and eclipses reached a remarkable degree of approximation (much above Græco-Arab computations) to the figures in Laplace's Tables.[59]

The Hindus were acquainted with Greek astronomy and its merits. Varaha-mihira's (A.D. 505–587) candid acknowledgment of the fact that this science is "well established" among the "barbarian" Yavanas (Ionians, i.e., Greeks) leaves no doubt on the point. The only question is about the amount and period of influence.

According to Burgess there was "very little astronomical borrowing between the Hindus and the Greeks." It is difficult to see precisely what the Hindus borrowed, "since in no case do the numerical data and results in the system of the two peoples exactly correspond."

A certain amount of foreign help may have given an impetus to the science in India. But the loan was thoroughly Hinduized. According to Whitney,[3] the Indians assimilated the Greek astronomy by

(1) The substitution of sines for chords, and

(2) The general substitution of an arithmetical for a geometrical form.

On the strength of subsequent developments, Seal claims that Hindu astronomy was not less advanced than that of Tycho Brahe (1546–1601).

Werner quotes passages to indicate that Hindu astronomical instruments were introduced into China. According to Mikami, Hindu astronomers served the Chinese government on the Astronomical Board, sometimes even as President (seventh century and after). Chinese translations of Sanskrit works * like " The Brahman Heavenly Theory " are also recorded. Several calendars were modelled on the Hindu, e.g., probably the one by Itsing (683–727). During the eighth century Hindu astronomy was introduced among the Saracens, also, as noticed above.

* Four Hindu books on astronomy and three on mathematics are scheduled in Book XXXIV of the " Sui Shoo " or History of the Sui Dynasty (A.D. 589–618).

IX

PHYSICS

PLAYFAIR makes the following remarks with regard to Greek physics: " Nothing like the true system of natural philosophy was known to the ancients. There are nevertheless to be found in their writings many brilliant conceptions, several fortunate conjectures, and gleams of light, which were afterwards to be so generally diffused."

The same remarks may be made, generally speaking, about Hindu physics. Both in methodology and achievements it exhibits almost the same strength and limitations as the Greek. But probably the attempts of the Hindu physicists were more comprehensive, and more coordinated with investigations in other branches of knowledge than those of the Greeks.

Some hypothesis of nature, i.e., of matter and energy, constituted the positive basis of each of the principal schools of Hindu philosophy, including metaphysics. The idea of a real "natural philosophy" was never absent from the intellectual horizon even of those who believed that " the proper study of mankind is man." There was no system of thought without its own physico-chemical theory of atoms, its own "laws of nature," and so forth. The most idealistic school had thus its own "materialistic" background. And the method of investigation, if not fully that of Baconian "experimental" induction, was not less fruitful and "experiential" than that of Aristotelian speculative logic.

The problems in natural philosophy, which engaged the attention, more or less, of every thinker in India, were of the kind described below:

1. The theory of atoms and molecular combinations. It is generally associated with the name of Kanada, the founder of Vaishesika philosophy. He has therefore been called the Democritus of India. Strictly speaking, there were almost as many atomic theories as the schools of Hindu thought. One or two may be mentioned:

(1) Vaishesika system: Atoms cannot exist in an uncombined state in creation. "The doctrine of atomism did not take its rise in Greece, but in the East. It is found in the Indian philosophy. Kanada . . . could not believe matter to be infinitely divisible. (Fleming's "Vocabulary of Philosophy.")

(2) Jaina system: The atoms are not only infinitesimal, but also eternal and ultimate. Atomic linking, or the mutual attraction (or repulsion) of atoms in the formation of molecules was analyzed by Umasvati (A.D. 50) with a most remarkable effect. The Jainas hold that the different classes of ele-

mentary substances are all evolved from the same primordial atoms. "The intra-atomic forces which lead to the formation of chemical compounds do not therefore differ in kind from those that explain the original linking of atoms to form molecules."[59]

2. General properties of matter: These were analyzed and defined not only by Kanada and his school, but also by the Jainas, Buddhists, and other rivals and contemporaries. A few such concepts were elasticity, cohesiveness, impenetrability, viscosity, fluidity, porosity, etc. Capillary motion was illustrated by the ascent of the sap in plants from the root to the stem, and the penetrative diffusion of liquids in porous vessels. Upward conduction of water in pipes was explained by the pressure of air.[59]

3. The doctrine of motion: Motion was conceived in almost every school of thought as underlying the physical phenomena of sound, light, and heat. This motion was known to be not only molar and molecular, but also the subtle motion lodged in the atoms themselves, i.e., the very principle of matter-stuff.

4. Time and Space: In order to be precise and definite in their calculations the Hindus conceived infinitesimally small magnitudes of time and space. The instruments of measurement were crude. The attempt to distinguish from one another the varying grades of "least perceptible" sound, light, heat, time, etc., has therefore to be taken for what it is worth. An atom (*truti*) of time was regarded as equal to

$\frac{1}{33750}$th of a second. The thickness of the minimum visible (*trasa-renu*), e.g., the just perceptible mote in the sunbeam, was known to be $\frac{1}{349525}$th of an inch. The size of an atom was conceived to be less than $\pi.3.5^{-1}.2^{-62}$ of a cubic inch. "Curiously enough, this is fairly comparable (in order of magnitude) with the three latest determinations of the size of the hydrogen atom."[59] No unit of velocity seems to have been fixed upon. But average velocity was measured in accordance with the formula $v=\frac{s}{t}$. These measurements were not arbitrary poetic guess-works. It is on the basis of these that a remarkably accurate measurement of the relative pitch of musical tones was made, and the instantaneous motion of a planet determined (and thus the "principle" of the differential calculus discovered).

5. The doctrine of conservation: Both matter and energy were known to be indestructible. But though constant, they were known to be liable to addition and subtraction, growth and decay, i.e., to changes in collocation. This transformation was known to be going on constantly.

The following ideas about matter and energy may be gleaned from the writings of the Hindus.[59] Some of these should be regarded as real contributions to knowledge, though not demonstrated according to the modern methods:

(*a*) Heat:

(1) Light and heat were known to Kanada as different forms of the same substance.

(2) Solar heat was known to Udayana as the source of all the stores of heat.

(3) Heat and light rays were believed by Vachaspati (A.D. 850) to consist of very minute particles emitted rectilineally by the substances.

(4) Rarefaction in evaporation and the phenomenon of ebullition were correctly explained by Shamkara Mishra.

(*b*) Optics:

(1) The phenomena of translucency, opacity, shadows, etc., were explained by Udyotakara.

(2) The angle of incidence was known to be equal to the angle of reflection. This was known to the Greeks also.

(3) The phenomenon of refraction was known to Udyotakara.

(4) The chemical effects of light rays were known to Jayanta.

(5) Lens and mirrors of various kinds, spherical and oval, were used for purposes of demonstration. Light rays were focussed through a lens on a combustible like paper or straw. (The making and polishing of glass was a great industry in India. According to Pliny the best glass was that made by the Hindus.)

(*c*) Acoustics:

(1) Physical basis of sound: Two theories were held about the vehicle or medium of propagation. Shabara Swami knew it correctly to be the air. But Udyotakara and others knew it to be ether.

(2) Wave-motion: The sound-waves were under-

stood by both schools. But Prashastapada knew them to be transverse; and Udyotakara and Shabara Swami understood the transmission of sound to be of the nature of longitudinal waves.

(3) Echoes were analyzed by Vijnana-bhiksu.

(4) Sounds were distinguished according to their tones and over-tones, volume or massiveness, and quality or timbre, by Batsyayana, Udyotakara, and Vachaspati (*c* A.D. 850).

(5) Musical notes and intervals were analyzed and mathematically calculated in the treatises on music,[14] e.g., Sharamgadeva's "Samgita-ratnakara" ("Ocean of Music") (1210–47), Damodara's "Samgita-darpana" ("The mirror of music") (1560–1647), etc. The relative pitch of the notes of the diatonic scale was, according to Krishnaji Ballal Deval, accurately determined.

(6) The so-called Pythagorean law [12] of the vibration of stretched strings was known to the Hindus, viz., the number of vibrations (pitch of a note) varies inversely as the length of the string.[10]

(7) The Hindus knew that the octave above a note has twice as many vibrations as the note itself. They had thus arrived at the octave on which modern Eur-American music is based.

(*d*) Magnetism:

(1) Elementary magnetic phenomena could not but be observed. The attraction of grass, straw, etc., by amber, and the movement of the iron needle towards the magnet, were explained by Shamkara Mishra as due to *adrista*, i.e., unknown cause.

(2) Bhoja (*c* A.D. 1050) in his directions for ship-

building gave the warning that no iron should be used in holding or joining together the planks of bottoms intended to be sea-going vessels. The fear was entertained lest the iron should expose the ships to the influence of magnetic rocks in the sea, or bring them within a magnetic field and so lead them to risks.[34]

(3) Mariner's compass: Mookerji points out a compass on one of the ships in which the Hindus of the early Christian era sailed out to colonize Java and other islands in the Indian Ocean. The Hindu compass was an iron fish (called in Sanskrit *matsya-yantra* or fish machine). It floated in a vessel of oil and pointed to the north.

(*e*) Electricity: Most rudimentary electrical phenomena may have been noticed by Umasvati (A.D. 50). His theory of atomic linking was based on the idea that the two atoms to be combined must have two opposite qualities. He believed that atoms attracted and repelled each other according as they were heterogeneous (i.e., unlike), and homogeneous (i.e., like), respectively.[59]

X

CHEMISTRY

BOTH in the East and the West chemistry was at first alchemy. It was principally a handmaid to the science or art of medicine, and subsidiarily allied to metallurgy and the industrial arts. Whatever be the worth of that chemistry according to the modern standard, the Hindu investigators could give points to their European peers. They were, besides, teachers of the Saracens and of the Chinese.

Leaving aside the chemists or druggists in the medical schools of India, two great specialists in chemistry *as such* were Patanjali (second century B.C.) and Nagarjuna (early Christian era). Patanjali was also a philologist, his commentary on the famous grammar of Panini is well known. His "science of iron" (*loha-shastra*) was a pioneer work on metallurgy. Nagarjuna's genius also was versatile. He is the patron-saint of alchemists. He is credited with having founded or rather systematized the philosophy of *rasa* (mercury).

Some of the achievements of the Hindu brain have been genuine contributions to chemical science.

The Hindu chemical investigators of the fifth and sixth centuries A.D. (the age of Gupta-Vikramadityan Renaissance) were far in advance of Roger Bacon (thirteenth century). In fact, they anticipated by one millennium the work of Paracelsus (sixteenth century) and Libavius (seventeenth century). "The physico-chemical theories as to combustion, heat, chemical affinity were clearer, more rational, and more original than those of Van Helmont or Stahl." (Seal.)[44]

1. According to Prafullachandra Ray, the earliest Hindus knew of the distinction between green and blue vitriol. But Dioscorides, the Greek, and Pliny, the Roman, both belonging to the first century A.D., confounded the two. Even Agricola's ideas were not clear (1494–1555).

2. The scientific pharmacy of Sushruta was almost modern. About the preparation of caustic alkali he was careful enough to give the special direction that the strong lye is to be preserved in an iron vessel. It was far superior to the process of a Greek writer of the eleventh century who has been eulogized by Berthelot.[44]

3. According to Royle, the process of distillation was discovered by the Hindus.

4. By the sixth century the Hindu chemists were masters of the chemical processes of calcination, distillation, sublimation, steaming, fixation, etc.[59]

5. These processes were used by researchers of the Patanjali and Nagarjuna cycles in order to

bring about chemical composition and decomposition, e. g.,

(*a*) In the preparation of
   (1) Perchloride of mercury;
   (2) Sulphide of mercury;
   (3) Vermilion from lead, etc.
(*b*) In the extraction of
   (1) Copper from sulphate of copper;
   (2) Zinc from calamine;
   (3) Copper from pyrites, etc.

6. The importance of the apparatus in chemical research is thus described in "Rasarnava," a work on chemistry, of the eleventh century:

"For killing (oxidizing) and coloring mercury, an apparatus is indeed a power. Without the use of herbs and drugs, mercury can be killed with the aid of an apparatus alone; hence an expert must not disparage the efficacy of the apparatus." [44]

With this preamble the author introduced his account of the chemical laboratory, instruments, crucibles, etc.

7. In "Madanapala-nighantu," a work on drugs (*c.* 1374), zinc was distinctly mentioned as a separate metal. Paracelsus (1493–1541) was thus anticipated in India by about two hundred years.

8. The philosophy of mercury was a recognized branch of learning during the fourteenth century. It was one of the celebrated sixteen in Madhavacharya's collection of philosophical systems (1331).

He mentioned " Rasarnava " as a standard work on mercury.

9. " Rasa-ratna-samuchchaya " (treatise on mercury and metals) is a comprehensive work of the fourteenth century. It embodies practically the whole chemical, mineralogical, and metallurgical knowledge of the Hindus developed through the ages. Like the " Brihat Samhita " (sixth century A.D.) by Varaha-mihira, it is a scientific encyclopædia. It is specially remarkable for its section on the laboratory, directions for experiments, and description of apparatus.

10. The Hindus had no knowledge of mineral acids for a long period. But this defect was made up by their use of *Vid*, which, says Ray, could " kill all metals." This was a mixture containing *aqua regia* and other mineral acids *in potentia.* The substance had probably been discovered by Patanjali.[59] Mineral acids were discovered almost simultaneously both in India and Europe' during the sixteenth century.

The debt of Europe to Saracen chemistry or alchemy is generally acknowledged by historians of science.[68] This implies also Europe's debt to the Hindus; for they had taught these teachers of mediæval Europe.

Gebir, the earliest Saracen (Spanish) chemist (eighth century), was familiar with Hindu *rasayana* (alchemy and metallurgy, the seventh division of the " science of life," called *Ayur-veda*). He called carbonate of soda "sagimen vitri" from the Hindu name *sajji matti*. He also knew *tutia*, the Hindu name of copper sulphate.[78]

The Saracens themselves admitted their discipleship to the Hindu professors of medicine. Chemistry naturally passed along with the medical science from India into the Saracen Empire. The famous Arabic encyclopædia, "Kitaba al Fihrist," by Nadim (*c* 950) distinctly mentions the translation of Hindu medical works into Arabic under the patronage of Caliphs from Mansur to Mamun (*c* A.D. 750–850). Saracen scholars[51] of the thirteenth century, e.g., Haji Khalifa, also acknowledged what their predecessors had learned from the schools of Hindu medicine.

The medical practice in China also was considerably influenced by Hindu chemistry. In Book XXXIV of the "Sui Shoo" (A.D. 589–618) Nagarjuna appears under the Chinese appellation of Loong (i.e., Dragon) Shoo (i.e., Tree), the exact equivalent of the Hindu name, as an authority on recipes. The chemistry of the Tantrists was further assimilated by the Chinese through treatises like the "book of recipes narrated by Chipo" (Chinese for Shiva, the god of the "Tantras").

The history of science requires, therefore, a revision, in the department of chemistry as in algebra, arithmetic, etc., in the light of facts from the Hindu angle.

XI

METALLURGY AND CHEMICAL ARTS

INDIA was the greatest "industrial power" of antiquity. It was the manufactures of the Hindus, which, backed up by their commercial enterprise, served as standing advertisements of India in Egypt, Babylonia, Judæa, Persia, etc. To the Romans of the Imperial epoch and the Europeans of the Middle Ages, also, the Hindus were noted chiefly as a nation of industrial experts.

Some of the arts [52] for which the people of India have had traditional fame are those connected with (1) bleaching, (2) dyeing, (3) calico-printing, (4) tanning, (5) soap-making, (6) glass-making, (7) manufacture of steel, (8) gun-powder and fire-works, and (9) preparation of cements. All these imply a knowledge of industrial chemistry.

1. Patanjali, the founder of Hindu metallurgy, (second century B.C.) gave elaborate directions for many metallurgic and chemical processes, especially the preparation of metallic salts, alloys, amalgams, etc., and the extraction, purification, and assaying of metals.[59]

2. During the fourth century the Hindus could forge a bar of iron, says Fergusson, "larger than any that have been forged in Europe up to a very late date, and not frequently even now." [44]

3. Gun-powder "may have been introduced into China from India" about the fifth or sixth century A.D. (Journal of the North China Branch of R. A. S., New Series, vi. 82).

4. The secret of manufacturing the so-called Damascus blades was learned by the Saracens from the Persians, who had mastered it from the Hindus.[52] In Persia, the Indian sword was proverbially the best sword, and the phrase *jawabee hind* ("Indian answer") meant "a cut with the sword made of Indian steel."

5. During the sixth century [59] the Hindu chemists could prepare:

(1) Fixed or coagulated mercury;
(2) A chemical powder, the inhalation of which would bring on sleep or stupor;
(3) A chemically prepared stick or wick for producing light without fire;
(4) A powder, which, like anæsthetic drugs or curare, paralyzes sensory and motor organs.

6. The horticulturists of the same period were familiar with several mixtures and infusions, probably struck upon empirically, for supplying the requisite nitrogen compounds, phosphates, etc., to plants.

7. The metallurgists of the same period were familiar with the processes of extraction, purifica-

tion, "killing" (formation of oxides, chlorides, and oxy-chlorides), calcination, incineration, powdering, solution, distillation, precipitation, rinsing, drying, melting, casting, filing, etc. With the help of apparatus and reagents they subjected each of the known minerals to all these processes. Heat was applied in different measures for different ends.[59]

8. So early as the sixth century the mercurial operations alone were nineteen in number.

Pliny, the Roman of the first century A.D., noticed the industrial position of the Hindus as paramount in the world. India maintained the same position even in the seventeenth and eighteenth centuries, when the modern European nations began to come into intimate touch with her. This long-standing industrial hegemony of the Hindus was due to their capacity for harnessing the energies of Nature to minister to the well-being of man. They made several important discoveries in chemical technology. These may be generalized [59] into three:

(1) The preparation of fast dyes;
(2) The extraction of the principle of indigotin from the indigo by a process, which, though crude, is essentially an anticipation of modern chemical methods;
(3) The tempering of steel.

XII

MEDICINE

SUPERSTITIONS die hard. The progress of rationalism is slow. Hippocrates and Galen held a knowledge of astronomy or rather astrology to be essential to physicians. In Europe, even so late as the fifteenth and sixteenth centuries, diseases were regarded as punishments of God, and the intervention of priests was requisitioned where one should call on a physician or a surgeon.[40] Thus, when after the return of Columbus's party from the newly discovered America to the Old World, venereal diseases created havoc in every country in Europe, people used to offer masses and prayers and alms to assuage the wrath of God. From the Popes and Cardinals down to the soldiers and traders, every rank of the society was infected by the disease. It was, therefore, considered to be a visitation from heaven to punish the licentious and rectify the universal ribaldry of the times.

In fact, the pseudo-science of Galen (second century A.D.) continued long to be an incubus upon medical theory and practice in Europe. Absurd formulæ held the ground in the Christian pharmacopæas of continental Europe down to comparatively modern times. And the age of talismans, amulets, the fetish of royal touch, etc., is yet fresh in human memory. Really scientific medicine is very recent.[29]

It is in this perspective of the history of medicine that Hindu contributions to its science and art have to be read. Hindu achievements in this field as in others have not only an "historical" importance, but have some "absolute" value also. Besides, from the standpoint of comparative chronology, Hindu medicine has been ahead of the European and has been of service in its growth and development.

Two great names in Hindu medicine are Charaka (*c* sixth to fourth century B.C. ?), the physician, and Sushruta (early Christian era), the surgeon. They were not the founders of their respective sciences, but the premier organizers of the cumulative experience of previous centuries. In observation lay their great strength, the "natural history of disease" was their special study. Both these schools were in existence about 500 B.C. according to Hoernle. By the first and second centuries A.D. surgery was a well developed art. Many instruments were devised, of which 127 are mentioned. The materia medica grew from age to age with the introduction of new drugs (vegetable, animal and mineral), of which the therapeutic effects were tested by the "experiments" of researchers.

(1) The Hindus have had hospitals and dispensaries since at least the third century B.C. Asoka the Great was an educator and propagandist. Through his Rock Inscriptions he popularized, among other things, some of the more common medical recipes for the treatment of both men and animals. (The first Christian hospital was built in the fourth century A.D. under Constantine.)

(2) The smoking of datura leaves in asthma, treatment of paralysis and dyspepsia by nux vomica, use of croton tiglium, etc., are modern in Europe, but have come down in India since very old times.[51]

(3) The Hindus were the first in the world to advocate the "internal" use of mercury. Pliny knew only of its external use (first century A.D.). By the sixth century it was well established among Hindu practioners as an aphrodisiac and tonic. It is mentioned by Varaha-mihira along with iron (587).[44]

(4) The Greeks and Romans used metallic substances for external application. The Saracens are usually credited with their internal administration for the first time in the history of medicine. According to LeClerc, the first physicians in Europe, who used mercury, lived in the fifteenth century, and were induced to do so from reading the works of Mesue of Damascus (750).

But in this as in other matters the Hindus anticipated the Saracens, and, in fact, taught them. As Royle observes, the earliest of the Saracens had access to the writings of Charaka and Sushruta, who had given directions for the internal use of numerous metallic substances.

(5) In the prescriptions of Dr. Vagbhata mineral and natural salts had a conspicuous place. His book was translated into Arabic in the eighth century.

(6) From the sixth century on, every Hindu treatise on materia medica has more or less recommended metallic preparations for internal use. It was only after Paracelsus at the end of the sixteenth century that these had a recognized place in European science.[44]

Hindu medicine has influenced the medical systems of other peoples of the world.[51] The work of Indian physicians and pharmacologists was known in ancient Greece and Rome. The materia medica of the Hindus has influenced mediæval European practice also through the Saracens.

(1) Hippocrates (450 B.C.), the "father of medicine," was familiar with Hindu drugs. Thus he mentions pepper, cardamon, ginger, cinnamon, cassia, etc. Theophrastus (350 B.C.) mentions ficus indica and others among medicinal plants. Dioscorides (first century A.D.), the most celebrated compiler of Greek materia medica, mentions valeriana hardwickii, calamus aromaticus, etc. Ætius (fifth century) mentions collyrium indiarum, santalum, and other characteristic Hindu medicaments. Similarly Paulus Ægineta (seventh century) prescribes the internal use of steel, cloves, rhubarb, trypherum, etc. Pliny,

the Roman contemporary of Dioscorides, had also mentioned Indian medicinal plants and drugs.

The preparations of the Hindu pharmaceutical laboratories were thus in use in Greece as well as in the Hellenistic and Græco-Roman world. The Hindu inventions were bodily incorporated in the European systems. The Indian names, e.g., hardwickii, trypherum, etc., were retained; also, the original Hindu uses of the drugs. And all this before the age of Saracen intermediaries.[51]

(2) Hindu physicians were superintendents of Saracen hospitals at Bagdad. The introduction of Indian drugs among Moslems has been acknowledged by their own medical men.

Serapion, the earliest Saracen author of materia medica (eighth century), mentions the Hindu Charaka. So also his followers, Rhazes and Avicenna.[78]

The Saracen physicians were surprised at the boldness with which Hindu practitioners prescribed the internal use of powerful metallic drugs. "Taleef Shareef" (Playfair's translation) is quoted by Udoychand Dutt to indicate the Moslem admiration of the Hindu practice:

"White oxide of arsenic: The Hindu physicians find these drugs more effectual, . . . but I usually confine them to external application.

"Mercury: It is very generally used throughout India, . . . it is a dangerous drug.

"Iron: It is commonly used by physicians in

India, but my advice is to have as little to do with it as possible."

(3) The Chinese scholar-tourists studied Hindu medicine.* Itsing "made a successful study" of the subject while in India (671–95), though it was not his special mission.[64]

(4) The later Greek physicians, e.g., Actuarius (twelfth century), Myrepsus, etc., were influenced by Saracen doctors.[29] They used Hindu medicaments also. Thus like the pre-Saracen Paulus, Actuarius mentioned *tri-phala* or "three myrobalans." This traditional Hindu drug has a place in his materia medica under the name of "tryphera parva."

(5) The Persian (post-Caliphate) doctors of the fourteenth, fifteenth, sixteenth, and seventeenth centuries, also, made use of the original Sanskrit treatises as well as of the previous Arabic translations. Meer Mohammed Moomin has acknowledged his indebtedness to Hindu works in his "Materia Medica."[51]

* Nine *Polomen* (i.e., Brahman, or Hindu) books on medicine are mentioned in the bibliographical section (Book XXXIV) of the "Sui Shoo" (A.D. 589–618). Some of these books are complete in twenty-five volumes, and contain the "recipes of many masters."

References to the "Sui Shoo" are due to the investigations of Mr. T. Y. Leo, late of the Chinese Legation of Washington, D. C.

XIII

SURGERY

The ancient Hindu surgeons gave expression to the most modern views about the importance of their science. They declared:

"Surgery is the first and best of the medical sciences, less liable than any other to the fallacies of conjectural and inferential practices, pure in itself, perpetual in its applicability, the worthy produce of Heavens, and certain source of Fame."

These ideas were prevalent among the medical practitioners during the first centuries of the Christian era, when the investigations of the Sushruta-cycle were being organized into a system.

Another very remarkably modern idea of these surgeons was that "the first, best, and most important of all implements is the hand."[79]

Surgery is one of the oldest branches of medical science in India. The Hindu term for it is *Shalya* or the "art of removing foreign substances from the body, especially the arrow." It seems to have had its origin in warfare and in the accidents of outdoor work, e.g., hunting and agriculture.

The Hindu surgeons performed lithotomy, could extract the dead fœtus, and could remove external matter accidentally introduced into the body, e.g., iron, stones, hair, bones, wood, etc. They were used to paracentesis, thoracis, and abdominis, and treated different kinds of inflammation, abscesses, and other surgical diseases. Hazardous operations and the art of cutting, healing ulcers, setting bones, and the use of escharotics, were the *forte* of a section of India's medical men.

Dissection of the human body and venesection were normal facts in medical India. The doctors of the Sushruta school declared that dissection was necessary for a correct knowledge of the internal structure of the body. Dissection gave them an intimate knowledge of the diseases to which the body is liable. It also helped them in their surgical operations to avoid the vital parts.[79] It gave them, besides, an accurate knowledge of the human anatomy.[23]

The Hindu surgical laboratory consisted of at least 127 instruments. The operators were used to the manipulation of saws, lancets, needles, knives, scissors, hooks, pincers, probes, nippers, forceps, tongs, catheters, syringes, loadstones, rods, etc.

For laboratory practice students operated on wax, gourds, cucumbers, and other fruits. Tapping and puncturing were demonstrated on a leather bag of water or soft mud. Fresh hides of animals, or dead bodies, were used in the demonstration of scarification and bleeding. The use of the probe was practised on hollow bamboos. Flexible models of the human body were in use for practice in bandaging. Caustics and cauteries were used on animals.[78]

Lest one should smile over this primitive stage of the science, it is fair to remember the barber-surgeons of Europe in the fifteenth and sixteenth centuries.

One need, moreover, resist the temptation of comparing or contrasting this ancient Hindu surgical theory and practice with the marvels of modern surgery. By the side of the latest discoveries and inventions, any achievements of the human brain in the past, whether in the East or the West, are but children's toys. "So rapid has been our surgical progress that a Velpeau, a Sir William Ferguson or a Pancoast, all of whom died within the last thirty years, could not teach modern surgical principles nor perform a modern surgical operation; . . . Our modern operations on the brain, the chest, the abdomen, and the pelvis, would make him wonder whether we had lost all our senses, until seeing the almost uniform and almost painless recoveries, he would thank God for the magnificent progress of the last half-century, which had vouchsafed such magical, nay, almost divine, power to the surgeon."[74]

XIV

ANATOMY AND PHYSIOLOGY

HIPPOCRATES, the founder of Greek medicine, was unacquainted with anatomy and physiology. His ignorance was due to the superstitious respect which the Greeks paid to their dead.[29] But the fathers of Hindu medicine were remarkably accurate in some of their observations and descriptions.

The Hindus have described 500 muscles—400 in the extremities, 66 in the trunk, and 34 in the region above the clavicle. They knew of the ligaments, sutures, lymphatics, nerve plexuses, fascia, adipose tissue, vascular tissue, mucous membrane of the digestive canal, synovial membranes, etc.[28]

(*a*) *Osteology*

The anatomical system of the Hindus was almost modern. As Hoernle remarks: "Its extent and accuracy are surprising, when we allow for their early age—probably the sixth century B.C.—and their peculiar method of definition."

There are about 200 bones in the human body according to modern osteology. Charaka counted 360, and Sushruta 300. The former counted the 32 sockets of teeth and the 20 nails as separate bones. These were not admitted by Sushruta.

The additional 100 in Sushruta's count, however, has to be explained. This large excess is principally due to the fact that, like Charaka, he regarded the cartilages and the prominent parts of bones (the modern "processes" and "protuberances") as if they were separate bones.[23] In Europe the first *correct* description of the osseous system was given by Vesalius in 1543.

(*b*) *The Doctrine of Humors*

The physiology of humors, whatever its worth, is older in India than in Greece. At any rate, the Hindu and the Greek humoral pathologies are independent systems. Hippocrates counted four humors, viz., blood, bile, water, and phlegm; but Charaka propounded three, viz., air, bile, and phlegm.

(*c*) *Digestion*

The Hindu physicians knew the digestive system well and described it satisfactorily.

1. The function of different digestive fluids was understood. They were familiar with the acid gastric juice in the stomach. They knew also that in

the small intestines there is a digestive substance in the bile.

2. They were familiar with, and explained, the conversion of the semi-digested food (chyme) into chyle, and of that again into blood.

3. They explained the chemical changes by the action of metabolic heat.

(d) *Circulation of Blood*

In Europe, previous to Harvey's epoch-making discovery (1628), "the movement of the blood was believed to be confined to the veins, and was thought to be a to-and-from movement." (Halliburton).

The Hindus knew that the heart (1) receives the chyle-"essence," i.e., venous blood, (2) sends it down to the liver, where it is transformed into red blood, and (3) gets it back as red blood from the liver.

There was thus the idea of a *chakra* or wheel, i.e., self-returning circle of "circulation."[59]

But the Hindus did not understand the process clearly. (1) They did not know that the pathway of the blood round and round the body is a "double circle," i.e., "systemic" circulation and "pulmonary" circulation. (2) Neither Charaka nor Sushruta, therefore, understood the function of the lungs in the oxygenation of blood. This was not known to the ancients in Europe also, e.g., to Galen (A.D. 130).

The Harveyan Circulation was thus not anticipated

by the Hindus. The Hindu conception of the vascular system is given below:

(1) There are two classes of blood-conductors (i) *shira* or artery (?) and (ii) *dhamanee* or vein (?),

(2) The heart is connected with the liver by both,

(3) The *dhamanees* bring the impure blood (venous) from the heart into the liver, and the *shiras* conduct the pure (arterial) blood from the liver into the heart.

(*e*) *Nervous System*

Neither in India nor in Europe did the ancients understand the nervous system. Aristotle's error was committed by Charaka and Sushruta also. They all regarded the heart to be the central organ and seat of consciousness. The nerves (sensory and motor) were believed to ascend to and descend from the heart.

Later investigators, however, corrected this mistake both in the East and the West. Like Galen, the Greek (second century A.D.), the Tantrists and Yogaists of India came to know the truth that the brain (and the spinal cord) is the real organ of "mind."

According to Bamandas Basu the nervous system is more accurately described in the mystical "Tantras" than in purely medical treatises. We get the following from "Shiva Samhita:"[55]

1. Familiarity with the brain and spinal cord;
2. The idea that the central nervous system is composed of gray and white matters;

3. Familiarity with the lateral ventricles of the brain (through the fourth and third ventricles);
4. Familiarity with the ganglia and plexuses of the cerebro-spinal system;
5. The idea that the brain is composed of *chandra-kala* or convolutions resembling half-moons;
6. The idea that the six *chakras* are the vital and important sympathetic plexuses, presiding over all the functions of organic life. Yoga or contemplation means control over the functions of these plexuses.

According to Seal, also, the enumeration, by Yogaists, of the spinal nerves with the connected sympathetic chain and ganglia, is a distinct improvement on the anatomical knowledge of Charaka and Sushruta.[59] Thus, according to the Yoga physiologists,

(1) The *Susumna* is the central cord in the vertebral column. The two chains of sympathetic ganglia on the left and the right are named *Ida* and *Pingala* respectively. The sympathetic nerves have their main connection with *Susumna* at the solar plexus. There are 700 nerve-cords in the sympathetic-spinal system.

(2) The soul has its special seat within the *Brahmarandhra* above the foramen of Monro and the middle commissure, but traverses the whole cerebro-spinal axis, up and down, along the *Susumna*.

XV

EMBRYOLOGY

It is desirable at the outset to remember two facts in connection with modern embryology:

1. It is only in recent years, thanks to the most magnifying microscopes, that the science has made real progress through the study of cells (" cytology ").

2. Even Darwin believed that the children resemble their parents because the parents contribute multitudes of minute particles from their own tissues to form the cells of their offspring. But this theory of " pan-genesis " has been subsequently proved to be wrong.[67]

In the history of science Hindu embryologists deserve recognition (1) as having made precise observations, some of which are great approximations to the latest demonstrated truths, and (2) as having guessed at theories, some of which are eminently suggestive. As for pseudo-biological hypotheses, India has not been more prolific than Europe from Hippocrates to Buffon.[29]

Some of the facts observed and explained by Charaka and Sushruta are given below:

(1) All the members of the human organism are formed at the same time, but are extremely small, as the first sprig of the bamboo contains the leaves, etc., of the future plant.[79] This idea of the development of the fertilized ovum by "palingenesis" survived in India after a long struggle with rival theories. It is an established truth to-day that though we find cells of one type in glands, of another type in the brain, of another type in the blood, and so forth, nevertheless all of them sprang from one original single cell.[67]

(2) "The hard substances of the fœtus, as hairs, bones, nails, teeth, vessels, ligaments, etc., are produced from the semen, and resemble the same part as in the father; and the soft parts as flesh, blood, fat, marrow, heart, navel, liver, spleen, intestines, etc., are formed principally from the blood of the mother, and resemble her."[79] This is virtually the Darwinian pangenesis now exploded.

(3) Weisman's theory of "germinal continuity" is the greatest discovery of modern embryology. It is now held that "somatic" cells contribute absolutely nothing to the original germ-plasm, that no parent ever produces a germ cell, that the individual inherits nothing from his parents, but both he and they obtain their characteristics from a common source, and that the line of descent or inheritance is from germ-cell to germ-cell, not from parents.[67] This recent idea about the physical basis of inheritance brings out the distinction between germ-cells and body-cells

(somatic). It was guessed to a certain extent by the Hindu biologists also in their controversy regarding the transmission of congenital deformities and constitutional diseases of parents to offspring.

Atreya held that " the parental seed (germ-plasm) contains the whole parental organism in miniature (or *in potentia*), but it is independent of the parents' developed organs, and is not necessarily affected by their idosyncrasies or deformities." The germ-plasm was described as an organic whole independent of the developed parental body and its organs. The physiological characters and predispositions of the offspring were explained as being determined by the constituent elements of this parental seed. The continued identity of the germ-plasm from generation to generation may be taken as a corollary to this, though nowhere expressly stated.[59]

(4) Elementary facts about impregnation, the cycle of sex, menopause, etc., could not but be observed:

(*i*) The menses continue for seventeen days during which the woman may be impregnated, and not at any other season. This Hindu idea of absolute sterility after a certain number of days is still held by some modern physiologists, though not a demonstrated truth.

(*ii*) The menses remain till the fiftieth year, when the woman is of a weak constitution; but it continues longer when the individual is strong.

(5) Modern physiology would not reject the little

kernel of truth that there is in the following statement: "The menses, after conception, goes in part to form the placenta, and as the blood flows every month, it coagulates to form the embryo; an upper layer being added every month to the embryo; and another portion to the breasts, of the mother, by which the mammæ are increased in size."[79]

(6) The stages of fœtal development described on the basis of post-mortem operations and major operations in obstetric surgery had also much of the truth that has been established in recent years.

"In the first month the mixture of the semen and menses forms a small mass like a pea; seven days after conception, it has the form of a bubble or inflated bag. On the tenth it is red, and on the fifteenth it resembles a small round piece of flesh. At one month it has small fibres proceeding from it and is animated with life."[79]

One need not try to compare with this account the advanced and definite ideas of modern embryology about the development of the successive generations of cells, from the original fertilized ovum, e.g., "morula," "blastocyst," "yolk sac," "entoderm," "ectoderm," "mesoderm," etc. But the following may be accepted for "popular" purposes:

In the third month five eminences appear, which when developed become the hands, feet and head. In the sixth month all the members of the body are formed, etc.[79]

(7) The following observation about the develop-

ment of the rudimentary organs of reproduction contains a suggestive hint:

The fœtus for a time remains indeterminate, and then takes on a definite male or female character. In the second month the sexual character is indicated by the shape of the fœtus, the shape of a round joint (?) indicating the male sex, and the elongated shape, as of a muscle (?), the female sex.[59]

(8) What determines sex? Can sex be produced at will? These questions have engaged the attention of scientists as well as quacks all through the ages both in the East and the West. The following Hindu ideas have had their European duplicates:

(*i*) When conception occurs on the unequal days of menses, a female child will be born.

(*ii*) Should the germ have more of the qualities of the semen, a male child will be formed, and of the menses, a female child.[79]

(*iii*) Before the fœtus takes on a definite male or female character, the development of the sex may be modified to some extent by food and drugs.[59]

Modern scientists have advanced several theories about sex-determinants.[17] The truth remains yet to be discovered.

XVI

NATURAL HISTORY

MINERALS, plants, and animals were objects of study among the ancients and mediævals in India as in Europe. But nothing approaching the "sciences" of mineralogy, botany, and zoology was achieved anywhere.

The discovery of the microscope in 1683 is the real beginning of the study of plant and animal anatomies and of the internal structure of minerals. The birth of modern chemistry in the work of Priestley and Lavoisier at the end of the eighteenth century started the physiology of plants and animals as well as the determination of the composition and constitution of minerals. In 1809 exact measurements of crystalline forms of many minerals were made. The perfection of the microscope in 1867 has given a great impetus to all these sciences during the last half-century.[13]

All previous studies in minerals had been under the thraldom of alchemy. The researchers were swayed by mythological and metaphysical notions.[36] Roger Bacon believed that the "philosopher's stone" was able to transform a million times its weight of base metal into gold. It was no unusual assertion that the fortunate possessors of the "elixir of life" had been able to prolong their lives to 400 years and more.[31] Even Libavius (1616), who combated the excesses of Paracelsus and the employment of "secret remedies," believed in the transmutation of metals and the efficacy of potable gold.[44]

Studies in plant life from Theophrastus (370–286 B.C.), the "father of botany," down to the revival of learning in the sixteenth century were mere observations in agriculture, horticulture, forestry, pharmacy, etc.[20] The investigations regarding animals also did not go beyond the stage of "bionomics," i.e., the lore of the farmer, gardener, sportsman, and field-naturalist, including thremmatology or the science of breeding.[45]

In this "pre-scientific" mineralogy, botany, and zoology the Hindu students of natural history also played a part. Considerable power of observation was exhibited, as well as remarkable precision in description, and suggestiveness in expression. Their nature study was oriented to the practical needs of socio-economic life. It was minute and comprehensive, and so far as it went, avoided the fallacies of mal-observation and non-observation. Whatever be the value of the results achieved, the investigation was carried on in a genuine "scientific spirit."

HINDU ACHIEVEMENTS

(a) *Minerals*

The principal metals and gems were discovered, described, and utilized by the Hindus independently of any foreign help. In fact, in this branch of knowledge as in many others the people of India were the pioneers.

Mining has been in operation in India since the earliest times. The use of gems and precious stones as well as their identification,[50] also, have a long history among the Hindus.

1. The Hindus were the first to discover gold (Roscoe and Schorlemner).[55]
2. The Hindus taught the world the art of extracting iron from the ores (Roscoe and Schorlemner).[55]
3. Even in the Mosaic period (1491–50 B.C.) precious stones and gems were in use in India. (Ball).[55]
4. Homer mentions tin probably by its Sanskrit name *kastira*. (Birdwood.)[55]
5. The Hindus supplied gold to the Persian Empire in the fifth century B.C.; and the story of Indian "gold-digging ants" (miners) is famous in Greek literature through Herodotus and others.
6. At first the Hindus knew six metals; gold, silver, copper, iron, tin, and lead. They discovered zinc, the seventh metal, sometime during the fourteenth century. (It is mentioned by name as a separate metal in "Madanapala-nighantu," 1374). In Europe it was discovered by Paracelsus in 1540.
7. The Hindu "doctrine of seven metals" was not, like the Greek and Saracen, influenced by the doctrine of the mystic influence of the seven planets.[44]

8. Examination of the genuineness of gems was an art even in the first century B.C. (cf. "The Troy Cart," a drama by Shoodraka).
9. There have been different methods of enumeration and classification of the precious gems in different periods. The last important phase is embodied in the "doctrine of nine gems." These are ruby, pearl, coral, emerald, topaz, diamond, sapphire, gomeda (agate or zircon), and vaidurya (chrysoberyl, or lapis lazuli). This doctrine was enunciated probably in the tenth century by the astronomer Shreepati.[49]
10. The nine gems are believed to have a mystic connection with nine planets. Shreepati was the first to add Rahoo (personification of the ascending node of the moon) and Ketoo (moon's descending node) to the list of the generally recognized seven planets.[49]

(*b*) *Plants*

Scientific observation was applied to the phenomena of the vegetable kingdom. The body of knowledge arrived at through the colligation of facts consisted, however, in mere guesses or hints of truth.

The following ideas of rudimentary plant-physiology have been credited to the experience of the "rhizotomi," pharmocologists, plant-physicians (*Briksa-yurvedists*) and horticulturists of ancient and mediæval India by Bhimchandra Chatterji:

1. Sexuality: Flowers are the organs of plants.
2. Phosphorescence, and exudation of water.
3. Photo-synthesis: The sun is the source of energy in the fuel; (*i*) plants assimilate potential energy from the

sun, (*ii*) the less refractive rays (red, yellow, and orange) of the setting sun are specially adapted to assimilation by plants.

4. Plants are living organisms: They have among others the following phenomena of life: (*a*) sap-circulation, (*b*) power of movement, heliotropic, nyctitropic and other movements, sensitiveness to touch ("bashfulness") etc., (*c*) growth and reproduction.

Characteristics of plant life as known to the Doctors of Nyaya (logic) are thus given by Seal:

(1) Udayana (*c* A.D. 975) notices in plants the phenomena of life, death, sleep, waking, disease, drugging, transmission of specific characters by means of ova, movement towards what is favorable and away from what is unfavorable.

(2) Gunaratna (*c* A.D. 1350) enumerates the following: (*i*) stages of infancy, youth and age; (*ii*) regular growth; (*iii*) various kinds of movement or action connected with sleep, waking, expansion and contraction in response to touch; also movement towards a support or prop; (*iv*) withering on wound or laceration of organs; (*v*) assimilation of food according to the nature of the soil; (*vi*) growth or decay by assimilation of suitable or unsuitable food as prescribed in the science of the diseases of plants and their treatment (*Briksayurveda*); (*vii*) disease; (*viii*) recovery from diseases or wounds by the application of drugs; (*ix*) dryness, or the opposite, due to the sap which answers to the chyle (*rasa*) in animals; and (*x*) special food favorable to impregnation.

Various classifications of plants (into groups with subdivisions) were attempted. These were, like the system of Jussieus, mostly based on properties.[54] They were mainly useful hints for practical men interested in economic botany. Identification was thus rendered easier than in the system of the early European botanists, which, according to Sachs, was too vague and insufficient for the purpose.

(*c*) *Animals*

Animals have had an important place in the medicine, dietetics, economic life, fine arts and religion of the Hindus. The people have thus had experience of the life-habits, habitats, external characteristics, etc., of animals, both domestic and wild. This accounts for their intimate familiarity with the topics generally treated of in descriptive zoology.

1. Like the science of the diseases of plants, veterinary science also is very old in India. The Hindus had hospitals for animals in the third century B.C.

2. The Hindus could set fractures and dislocations in animals. They were perfectly acquainted with the anatomy of the goat, sheep, horse, and other animals used in sacrifices.[19]

3. They were specialists in the science of horses and elephants, the two animals important in warfare. Shalihotra is the founder of the science of horses, and Palakapya of the science of elephants. There is a vast literature on the subject.

4. Equine dentistry: The changes in the development and color of the six incisors of the lower jaw

constituted, in Hindu practice, the guide to the age of the horse. This is modern European practice also.

5. Snake-poison has been used as an article in Occidental materia medica during the last two or three decades. But it has been a recognized drug in India since early times.

6. The toxicologists of the Sushruta school of medicine devoted special attention to the study of snakes. That study was followed up in some of the "Purana" schools.

(*a*) Five different genera or families are described by Sushruta-Nagarjuna. Of these one is non-venomous, and the others are venomous. One of the venomous families is hybrid. The varieties of each are mentioned, as well as their longevity and other characteristics.

(*b*) The "Bhavisya Purana" records that the snakes (*Naiæ*) copulate in May or June, gestate during the rainy months that follow, and bring forth about 240 eggs in November. Most of these are devoured by the parents, but those that are left break forth from the shell in about two months. By the seventh day the young snakes turn dark; in a fortnight (or twenty days, according to another account) the teeth come out. The poison is formed in the fangs in three weeks, and becomes deadly in the twenty-fifth night. In six months the snakes shed the skin. The joints on the skin (scales or scutes) are 240 in number (perhaps the sub-caudals were not counted).[59]

7. Various systems of classification were built up: (1) according to the nature of generation, e.g., from placentalia, or egg, etc. (in the writings of the schools of medicine); (2) according to the habitat and mode of life, and usefulness to man; (3) according to the number of senses possessed by animals. (This was the system of Umasvati, A.D. 40).[59]

8. The Sushruta-school names (1) six varieties of ants, (2) six varieties of flies, (3) five varieties of mosquitoes (including one marine and one mountain kind), (4) eight varieties of centipedes, (5) thirty varieties of scorpions, (6) sixteen varieties of spiders.[59]

9. Leeches have been used by Hindu surgeons since very early times. Sushruta gives a detailed account of their varieties, habits, mode of application, etc. There are twelve varieties of leeches, six of which are venomous and six useful. The venomous are found near putrid fish or animals in foul water. The good are found in clear deep pools which contain water-lilies.[11]

10. Ladyayana is quoted by Dalvana, the commentator to Sushruta, as a great authority on insects and reptiles. According to this ancient specialist, the various forms of insects are to be distinguished from one another by the following marks: (1) Dottings, (2) wings, (3) pedal appendages, (4) mouth, with antennæ or nippers, (5) claws, (6) sharp, pointed hairs or filaments, (7) stings in the tail, (8) hymenopterous character, (9) humming or other noise, (10)

size, (11) structure of the body, (12) sexual organs (13) poison and its action on bodies.[59]

11. Dalvana's descriptions of deer and birds are precise and complete.

The zoological lore of the Hindus is thus in all respects a good document of their general scientific interest in the facts and phenomena of the objective world. And some of their classifications were not less remarkable than those of Aristotle.[26]

CONCLUSION

In conclusion, a few general remarks may be made with regard to the cultivation of exact sciences among the Hindus:

1. Like the Greeks, as Whewell admits, the Hindus also "felt the importunate curiosity with regard to the definite application of the idea of cause and effect to visible phenomena," "drew a strong line between a fabulous legend and a reason rendered," and "attempted to ascend to a natural cause by classing together phenomena of the same kind." (This scientific attitude of mind Whewell does not find in any non-Greek except the Hindu! He forgets altogether the claims of the Chinese.)

2. Epoch by epoch, Hindu scientific investigation was not more mixed up with metaphysics and superstitious hocus-pocus than the European. It enlisted in its service the devotion of hosts of "specialists" in succession. Their sole object was the discovery of the positive truths of the universe or the laws of nature, according to the lights of those days.

3. There thus grew up in India a vast amount of specialized scientific literature, each branch with its own technical terminology. The positive sciences of

the Hindus were not mere auxiliaries or handmaids to the "architectonic" science of *neeti* or *artha* (i.e., politics, economics, and sociology.) The sciences (*shastras*) on plant and animal life, veterinary topics, metals and gems, chemistry, surgery, embryology, anatomy, symptomology of diseases, arithmetic, algebra, astronomy, architecture, music (acoustics), etc., had independent status. Besides, like Pliny's "Natural History," there have been scientific encyclopædias in Sanskrit, e.g., the "Brihat Samhita" (sixth century A.D.).

4. Scientific investigation was not confined to any particular province of India or to any race or class of the Hindu population. It was a cooperative undertaking, a process of cumulative effort in intellectual advance. Thus, among the heroes of Hindu medicine,[55] Charaka (*c* 600 B.C.) belongs to the Punjab in the N.W., Sushruta (*c* 100 A.D.) is claimed by the Punjab as well as Benares in the Middle West, Vagbhata (*c* 700) belongs to Sindh (Western India), Vrinda (900) to the Deccan (Middle South), Chakrapani (1050) to Bengal (Eastern India), Sharamgadhara (1350) to Rajputana (Further West), Visnudeva (1350) to Vijayanagara (Extreme South), and Narahari (seventeenth century) is claimed by Kashmir (Extreme North) but belongs most probably to Maharastra (Western Coasts).

5. No one hypothesis or theory dominated Hindu thought in any age, or monopolized the researches of all investigators in successive epochs. The intel-

lectual universe of the Hindus was "pluralistic." There were different schools criticising, correcting, and modifying one another's inquiries.

The schools of abstract philosophy grew ultimately to sixteen in the time of Madhavacharya (1350), "though as a southerner," says Haraprasad Sastri "he omits the two Shaiva schools of Kashmir and puts the school of Buddhist philosophy into one." There were 15 different schools of grammar in the sixth century B.C., 10 different schools of politics and economics in the fourth century B.C., various schools of dramaturgy and dancing in the second century B.C., and also various schools of *Kama* or sexology about the same time.

The diversity of scientific doctrines in India may be illustrated by the differences of views regarding the nature of life. The Charvakas (materialists and sensationalists) held "that life (as well as consciousness) is a result of peculiar combinations of dead mater (or the four elements) in organic forms, even as the intoxicating property of spirituous liquors results from the fermentation of intoxicating rice and molasses." According to a second school (the Samkhya), life is neither a bio-mechanical force nor any mere mechanical motion resulting therefrom. It "is in reality a reflex activity, a resultant of the various *concurrent* activities of the sensori-motor, the emotional and the apperceptive reactions of the organism." A third school (the Vedantist) rejects both these doctrines. According to this, "sensations do not explain life. Life must be regarded as a separate principle . . . prior to the senses."[59]

Another illustration may be given from Hindu physics. This relates to the various hypotheses of Sound phenomena. One school held that the physical basis of audible sound is a specific quality of air, and that air-particles flow in currents in all directions. A second school, e.g., that of Shabara Swami held that it is not air-currents but air-waves, series of conjunctions and disjunctions of the air-particles or molecules, that constitute the sound physical. A third school held that the sound-wave has its substrate not in air but in ether. Further, Prashastapada held the hypothesis of transverse waves and was opposed by Udyotakara who held that of longitudinal waves.

6. The story of scientific investigation among the Hindus is thus, like that among other nations, the story of a growth and development in critical inquiry, sceptical attitude, and rationalism. Historically and statistically speaking, superstition has not had a deeper and more extensive hold on the Oriental intellect than on the Occidental.

INDEX

WORKS BY BENOY KUMAR SARKAR

The Folk-Element in Hindu Culture—A Contribution to Socio-Religious Studies in Hindu Folk-Institutions. With the assistance of H. K. Rakshit. Crown 8vo, $5 net.

The Science of History and the Hope of Mankind. Crown 8vo, $0.90 net.

Introduction to the Science of Education. Crown 8vo, $1.25 net.

Hindu Achievements in Exact Science. A Study in the History of Scientific Development. 12mo.

LONGMANS, GREEN & CO.

Shookra-neeti (Hindu Politics). Rendered into English from Sanskrit with Introduction and Notes. With an index by Narendranath Law, author of "Studies in Ancient Hindu Polity." 8vo, $3 net.

The Positive Background of Hindu Sociology. Book I. Non-Political. With appendices by Brajendranath Seal, King George V. Professor of Philosophy in the University of Calcutta. 8vo, $3.50 net.

CONTENTS:—Relativity of *Neeti-shastras* (Political Science)—Data of Ancient Indian Geography—Data of Ancient Indian Ethnology—Data of Ancient Indian Mineralogy—Data of Ancient Indian Botany—Data of Ancient Indian Zoology.

Chinese Religion through Hindu Eyes. A Study in the Tendencies of Asiatic Mentality—With an Introduction by Dr. Wu Ting-fang, Late Chinese Minister to Washington, D. C. $2.25.

Love in Hindu Literature. Price 75 cents.

CONTENTS:—Tagore—Kalidas—Vidyapati—Allegory of the Heart of a Woman—The Boiling-Point of Love-Heat—The Dignity of Sex—Sex-Element in Hindu Culture—Sex-Movements and Sex-Sciences—Sexual and Spiritual—Futurism of Young India.

PIANINI ACADEMY OF RESEARCH, ALLAHABAD, INDIA;
THE COMMERCIAL PRESS, SHANGHAI;
MARUZEN CO., TOKYO; LUZAC & CO., LONDON.

OTHER WORKS ON INDIA AND HINDU CULTURE

The Foundations of Indian Economics by Professor Radhakamal Mukerjee. Price $3 net.

The Positive Sciences of the Ancient Hindus by Professor Brajendranath Seal. Price $4.25 net.

A History of Indian Shipping by Professor Radhakumud Mookerji. Price $2.50 net.

Studies in Ancient Hindu Polity by Narendranath Law. Price $1.25 net.

The Arya Samaj by Lajpat Rai. Price $1.75 net.

The Promotion of Learning in India by Muhammadans by Narendranath Law. Price $4.50 net.

The Promotion of Learning in India by Early European Settlers by Narendranath Law. Price $1.50 net.

The Fundamental Unity of India by Radhakumud Mookerji. Price $1.25 net.

LONGMANS, GREEN & CO.

BENGALI WORKS BY BENOY KUMAR SARKAR

1. Lessons on Sanskrit. 2. Lessons on English. 3. The study of language. 4. History of education in ancient Greece. 5. Problems in education. 6. Studies in History. 7. Miscellaneous Essays. 8. The great leader of the Negro race (translation of Booker Washington's autobiography). 9. Translation of the historical section of Frederick List's "National System of Political Economy." 10. Trans-Himalayan Asia (Studies in Chinese Culture.) 11. The Modern World—a survey of present day tendencies in industry, education, literature, science, art, and social service—embodying the impressions of a tour round the world (1914–16) Six volumes—2500 pages.

POEMS. 1. **Asâdhya Sâdhan** (Doing the Impossible); (a) Personality, (b) We Bengalis, (c) Man Am I, (d) World's Ways, (e) Shakti-Yoga, or Energism. 2. **Dooniyâr Dâk** (The Call of the World): Sonnets on the lands and seas, and men and manners. 3. **Dooniyâ Âmâr** (The World is Mine): Translations from Chinese, Japanese, European and American poetry. 4. **Jeebaner Sâdâ** (Life's Responses): Studies in Situations and Attitudes.

This volume from the
Cornell University Library's
print collections was scanned on an
APT BookScan and converted
to JPEG 2000 format
by Kirtas Technologies, Inc.,
Victor, New York.
Color images scanned as 300 dpi
(uninterpolated), 24 bit image capture
and grayscale/bitonal scanned
at 300 dpi 24 bit color images
and converted to 300 dpi
(uninterpolated), 8 bit image capture.
All titles scanned cover to
cover and pages may include
marks, notations and other
marginalia present in the
original volume.

The original volume was digitized
with the generous support of the
Microsoft Corporation
in cooperation with the
Cornell University Library.

Cover design by Lou Robinson,
Nightwood Design.

Made in the USA
Monee, IL
13 September 2026

62976466R00065